Pushes and Pulls
On the Go
Ryan Smith
LIGHTBOX
openlightbox.com

LIGHTBOX

Go to
www.openlightbox.com
and enter this book's
unique code.

ACCESS CODE

LBXR8879

Lightbox is an all-inclusive digital solution for the teaching and learning of curriculum topics in an original, groundbreaking way. Lightbox is based on National Curriculum Standards.

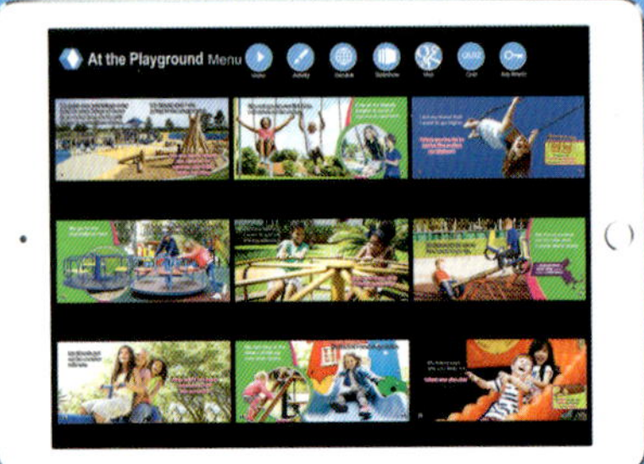

OPTIMIZED FOR
- ✓ TABLETS
- ✓ WHITEBOARDS
- ✓ COMPUTERS
- ✓ AND MUCH MORE!

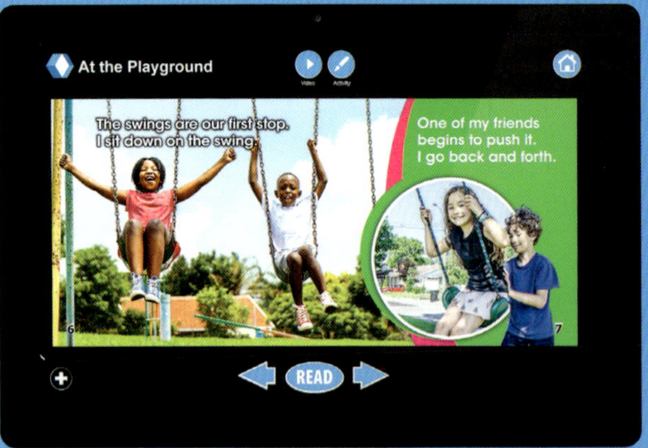

STANDARD FEATURES OF LIGHTBOX

 **AUDIO** High-quality narration using text-to-speech system

 **VIDEOS** Embedded high-definition video clips

 **ACTIVITIES** Printable PDFs that can be emailed and graded

 **WEBLINKS** Curated links to external, child-safe resources

 **SLIDESHOWS** Pictorial overviews of key concepts

 **INTERACTIVE MAPS** Interactive maps and aerial satellite imagery

 **QUIZZES** Ten multiple choice questions that are automatically graded and emailed for teacher assessment

 **KEY WORDS** Matching key concepts to their definitions

VIDEOS

WEBLINKS

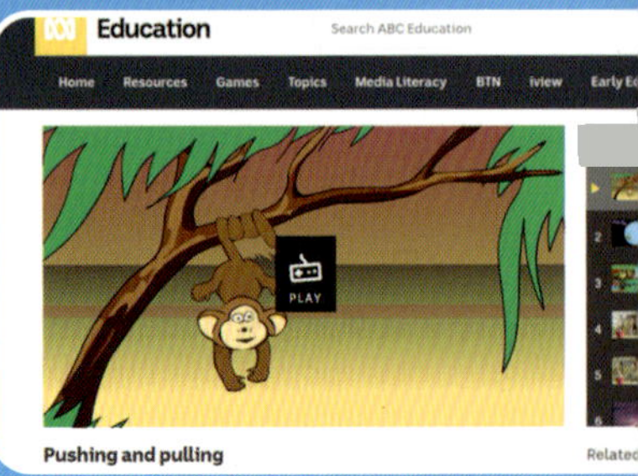

SLIDESHOWS

QUIZZES

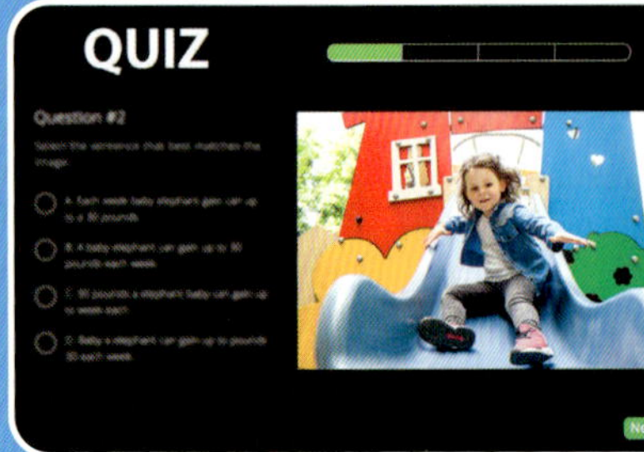

Pushes and Pulls

On the Go

Contents

We push and pull things every day. We push things to move them away from us. We pull things to move them closer.

Today, we are getting ready for a picnic in the park.

We can learn about the science of pushes and pulls on the go.

We drive to the store to shop for the picnic. I pull on the handle to open the car door.

I pull my seat belt across myself.

We are at the store.
I push the car door open.

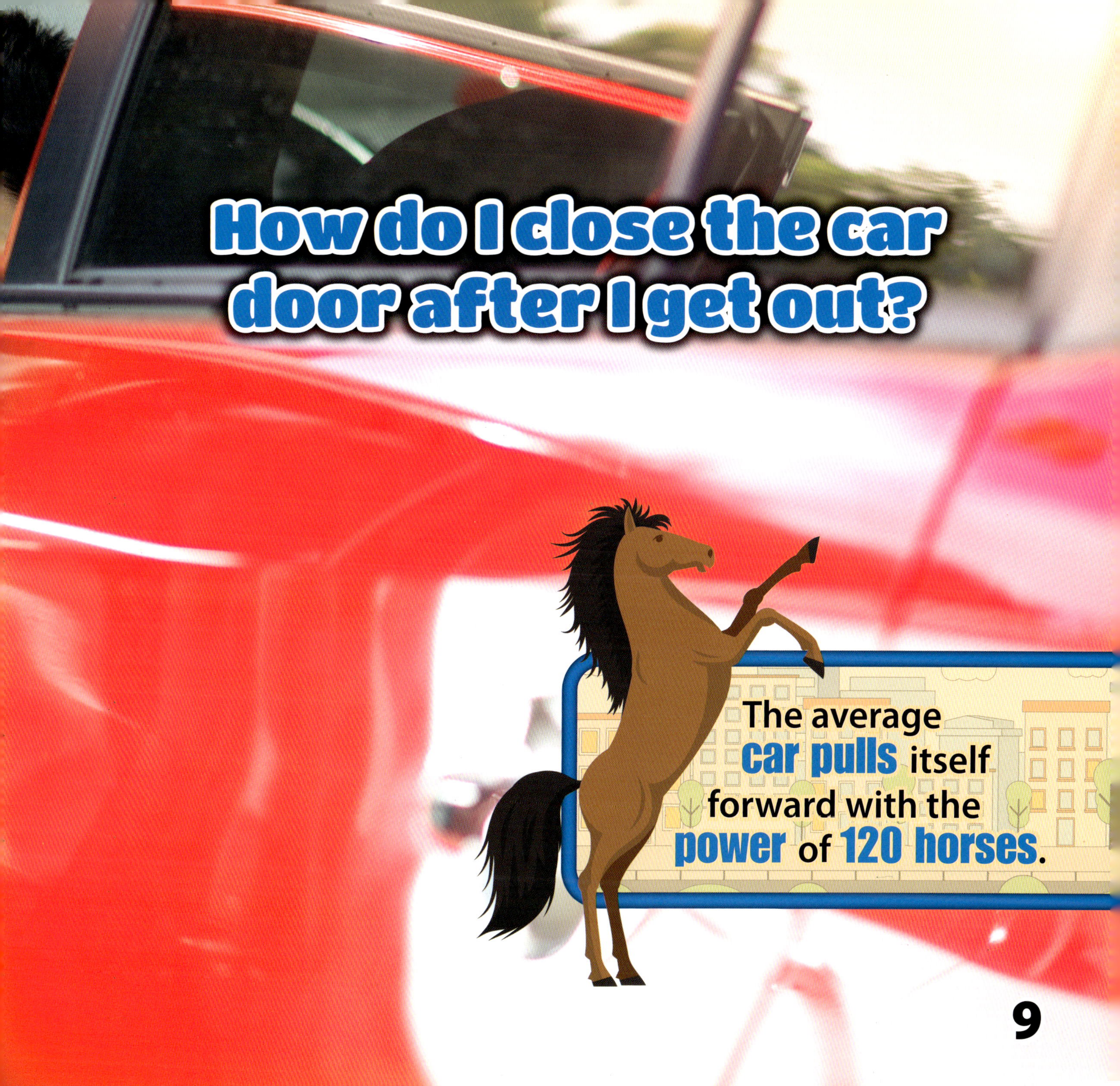

How do I close the car door after I get out?

The average **car pulls** itself forward with the **power** of **120 horses**.

At the store, we grab a shopping cart.

I push the cart through the aisles.

It gets harder to push the cart as we put things in it.

Why is it harder to push the cart?

I bring my wagon to the park.

It is easy to pull my siblings in my wagon.

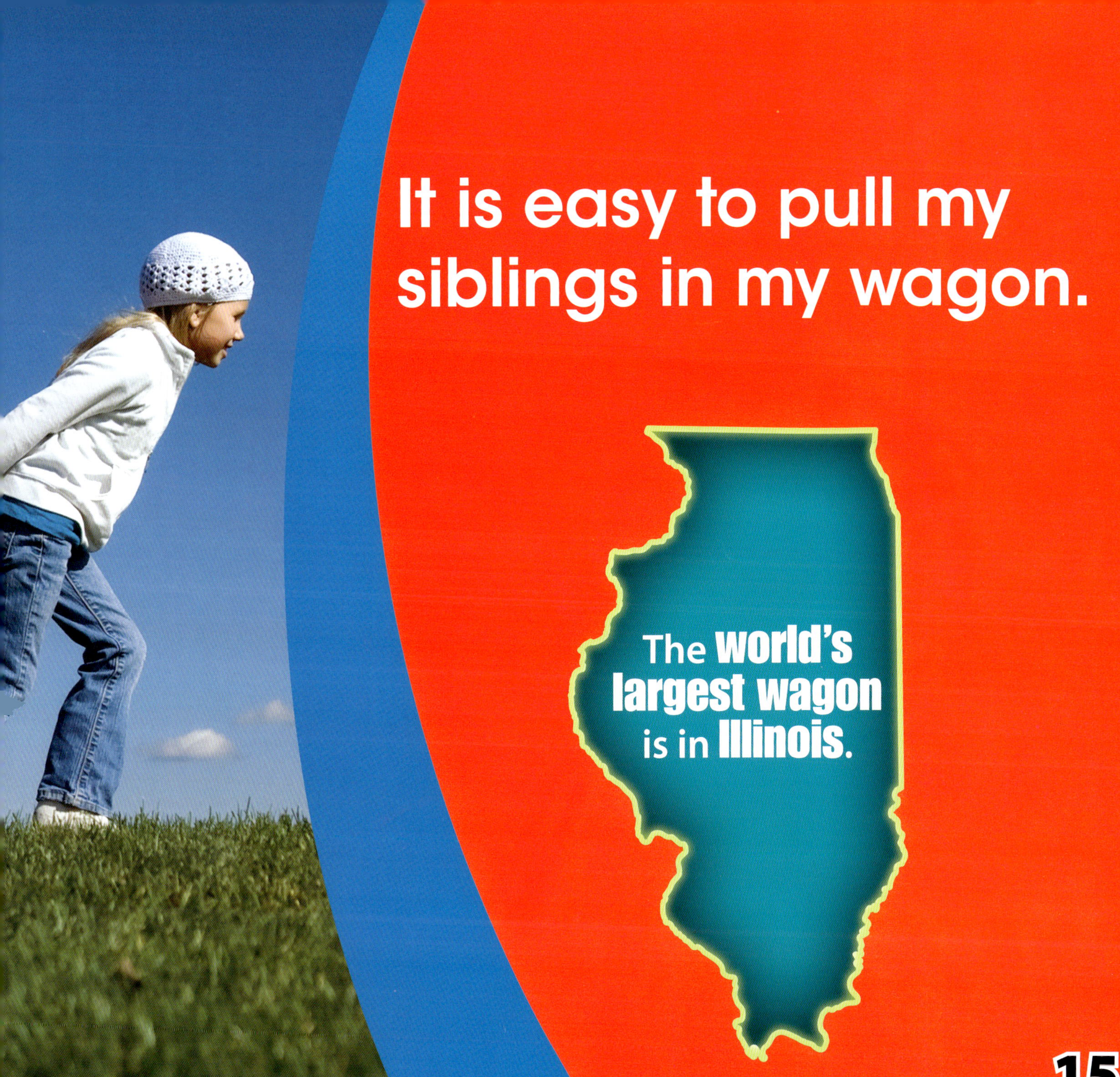

Our picnic spot is on top of a hill.

Will it be easy
to pull the wagon
up the hill?

After lunch,
I fly my kite.

The wind pulls my kite
into the air.

The wind pulls my kite higher and higher.

How can I make my kite fly lower?

The **highest** a kite has ever flown is higher than **12 Empire State Buildings**.

See what you have learned about pushes and pulls on the go.

What do you push while on the go?
What do you pull while on the go?

KEY WORDS

Research has shown that as much as 65 percent of all written material published in English is made up of 300 words. These 300 words cannot be taught using pictures or learned by sounding them out. They must be recognized by sight. This book contains 50 common sight words to help young readers improve their reading fluency and comprehension. This book also teaches young readers several important content words, such as proper nouns. These words are paired with pictures to aid in learning and improve understanding.

Page	Sight Words First Appearance
4	and, away, day, every, from, move, them, things, to, us, we
5	a, about, are, can, for, go, in, learn, of, on, the
6	car, I, open
7	my
8	at
9	after, close, do, get, how, out, with
11	through
12	as, it, put
13	is, why
15	world
16	our
17	be, up, will
19	air, into
20	make
21	has, than

Page	Content Words First Appearance
5	park, picnic, pulls, pushes, science
6	door, door handle, store
7	seat belt
9	horses, power
10	shopping cart
11	aisles
14	wagon
15	Illinois, siblings
16	hill, spot, top
18	lunch, kite
19	wind
21	Empire State Building

Published by Smartbook Media Inc.
14 Penn Plaza, 9th Floor New York, NY 10122
Website: www.openlightbox.com

Library of Congress Control Number: 2020014811

ISBN 978-1-5105-5466-5 (hardcover)
ISBN 978-1-5105-5467-2 (multi-user eBook)

Printed in Guangzhou, China
1 2 3 4 5 6 7 8 9 0 24 23 22 21 20

052020
110819

Designer: Ana María Vidal Project Coordinator: Ryan Smith

The publishers acknowledges Getty Images, iStock, and Shutterstock as the primary image suppliers for this title.